GENUINE PLASTIC
RADIOS
OF THE MID-CENTURY

WITH VALUES

Ken Jupp & Leslie Piña

Schiffer Publishing Ltd

4880 Lower Valley Road, Atglen, PA 19310

Dedication

For Ken's daughters Nicole and Natalie for putting up with him and this project (because they are not teenagers yet).

For Leslie's husband Ramōn for his patience and help, (but not for Ravi until he is finished being a teenager).

And for radio collectors everywhere.

Book design by Leslie Piña
Layout by Blair Loughrey

ISBN: 0-7643-0108-X
Printed in Hong Kong
1 2 3 4

Published by Schiffer Publishing Ltd.
4880 Lower Valley Road
Atglen, PA 19310
Phone: (610) 593-1777; Fax: (610) 593-2002
E-mail: Schifferbk@aol.com
Please write for a free catalog.
This book may be purchased from the publisher.
Include $3.95 for shipping.

Please try your bookstore first.

We are interested in hearing from authors
with book ideas on related subjects.

Contents

Acknowledgements   **4**

Forward by Barry Gould   **5**

Introduction   **6**

The Radios: **12**

    Radios without clocks   12

    Clock radios   124

    Portable radios   164

    Transistors   188

Logos   **190**

    Motorola and Zenith Highlights   211

Bibliography   **213**

Index   **214**

Acknowledgements

Since color and plastic are themes of this book, it seemed appropriate to use colorful plastic backgrounds for the pictures. Wilsonart International, a major manufacturer and distributor of colorful plastic laminate, was making this versatile product at mid-century, and many radios were placed on countertops fabricated from it. Today, with an interest in fifties style, Wilsonart's color palette includes colors and patterns with a fifties flavor. We selected some to use with the radios, and the Wilsonart color name and number is included below. We would like to thank Wilsonart International, Inc. for their support of this project and for the use of decorative laminates for all of the backgrounds of radios photographed in this book. The following patterns were used:

The colors shown are printed representations. For maximum fidelity, please request product samples: 1-800- 433-3222

Thanks also to John Taylor of Zenith Electronics Corporation for providing vintage photographs and information, and to Mary Edith Arnold of the Motorola Museum of Electronics for providing information. Many of the radios pictured are from the author's (Ken's) collection, and we would like to thank his friends and relatives (his brother Dave also helped with the cover) who have given him radios over the years, and his parents Bob and Darlene Jupp for passing on their collecting genes (Darlene collects lamp finials; Bob collects wooden-handled monkey wrenches). Other radios were lent by collectors and dealers including: Bill Igo of Vintage Treasures in Columbus, Brett and Lori Gundlach, Rae Cavanaugh, Steve Dotts, Paul Oravetz, Tim Poling, Jack and Sylvia Schenkel, Jim Vattar, John Wichman, and Barry Gould. A big thanks to Barry for help with pricing and for writing the Foreword. Special thanks to Todd Vogelsinger of WILSONART International, Inc. and to the Central Ohio Antique Radio Association (C.O.A.R.A.). Ramón, as usual, assisted the author (Leslie) with the photography and already knows how much he is appreciated. Thanks again to Paula Ockner for proofreading, and any remaining errors are ours. Nancy and Peter Schiffer and Douglas Congdon-Martin were great to work with, as they always are. It was a fun project. Thank you all.

The authors welcome hearing from readers with questions, comments, or information about plastic radios. Please contact Ken at MANTISMARK@aol.com.

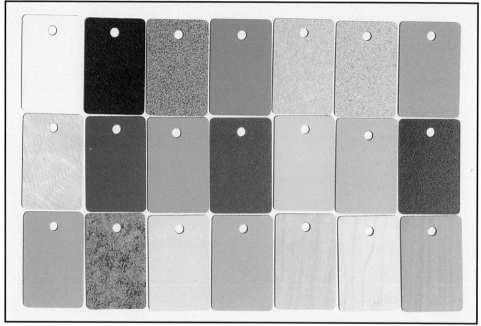

Top Row		Center Row		Bottom Row	
Alabaster	D431-60	Klondike	4512-60	Rosemist	D344-60
Black	1595	Lapis Blue	D417-60	Scopia	4556-60
Dakota Ridge	4557-60	Limerick	D454-60	Sundance Pink	D375-60
Emerald	D365-60	Mandarin	1511-60	Wisteria	D347-60
Erin Glen	4305-90	Marigold	D341-60	Fusion Maple	7909-60
Green Glace	4144-60	Nile Green	D362-60	Manitoba Maple	7911-60
Island Sky	D415-60	Port	D14-60	Natural Pear	7061-60

Foreword

Over the years, radio manufacturing has gone through many changes. Production before World War II focused on various styles of richly grained wood table and counsole models. There were also early plastic table radios of Bakelite and Catalin, but it was not until after the war that the production of plastic radios accelerated.

A new type of daring and exciting designs emerged at companies like Crosley, Emerson, Motorola, Zenith, and others. Designers and the companies that commissioned them took risks, and most paid off. By the early to mid 'fifties, designs of brightly colored table radios were downright fun.

Until recently, these mid-century plastic radios were often overlooked by serious collectors. The purpose of this book is to acquaint some and to re-acquaint others with a sample of these models that once delighted listeners of this past era. Those of us who listened at mid-century remember the AM band that brought news and the tunes of Elvis, the Beach Boys, and even the early Beatles. We went from a "big family radio" to a radio for any, perhaps every, room. Colorful plastic table models were found on kitchen counters, bedside tables, in garages, and the new "family rooms." These little five or six-tube wonders filled the family home with sound and with color.

As we approach the end of the century and the millennium, most of these radios are over forty years old, and they are getting harder to find in good condition. In addition, a new breed of radio collector has arrived -- with a passion for plastic. Where for years collectors of wood radios passed up the plastics, the reverse is happening today. Many "plasticholics" hardly take notice of the bulky wood units that once occupied the attention of most radio collectors. This new collector seeks out beautifully designed units for home and office shelves. Some insist that the radio is in working condition, because the old sound is part of the fun. Others just want the design and the color. Nostalgia is a force that drives many collectors, but there is a growing number who still appreciate the style though they were never exposed to the era. This is not surprising, when one compares plastic radios to the unimaginative units available today.

Leslie and Ken have extensive backgrounds in design and color. They have taken the time to seek out and (with Ramón's help) to photograph many representative examples of mid-century tabletops...we all hope you enjoy.

Barry Gould

Columbus, Ohio

Introduction

Collecting usually begins with the early items in a category and gradually evolves to include examples that were once considered to be ordinary or unworthy. Radios are no exception. Large console models and early table radios of exotic plastics like Bakelite or Catalin have been the focus of collections and literature. Specialties -- radios of single manufacturing companies or items such as transistors -- have also interested collectors and authors.

This book will focus on the "brittle" and brightly colored plastic radios of mid-century. Common mid-century radios made primarily of brightly-colored plastic represent a relative newcomer to the radio collecting arena. With designs that often resembled contemporary automobile grille and headlight configurations, these affordable radios were made by the major electronics companies and dozens of lesser-known manufacturers.

Not long ago, these sets would be the last to be sold at radio swap meets or auctions. Today, they are among the first to be scooped up by a new group of collectors and by those whose tastes have changed.

We have attempted to provide an overview of these genuine plastic radios at mid-century. It is not meant to be comprehensive. There were so many different models made by so many different companies that it would be unrealistic, if not impossible, to cover all of them. The order of photographs in this

Example of Catalin Radio
Fada. 1000. 1945. Maroon, yellow. **$600**

Catalin Corporation magazine advertisement.

Plaskon Company magazine advertisement.

book is divided into three roughly chronological parts: 1) radios without clocks, 2) clock radios, and 3) portables. Pocket transistors are touched upon at the end.

These examples of American industrial design and popular culture were once plentiful, but today they are elusive. As functional items nearly all were used, and they eventually lost knobs, gained cracks in cases or dial faces, or the brittle plastic chipped off corners and edges. Many were discarded or just packed away in attics when they stopped working, because as consumer items, plastic radios were considered disposable.

Like other popular collectibles, mid-century genuine plastic radios are both common and rare: common in poor condition and rare in perfect condition, especially in original boxes. Collectors naturally seek, but do not often find, the rare and

mint examples, so tolerance for minor imperfections is expected. Those with extraordinary design and color are among the most desirable, but even the less flamboyant radios have their own appeal. Even in less-than-perfect condition, they represent a period in history associated with innocence and fun; they bring on a certain nostalgia for the time, even for those who never knew it personally.

Though there are categories of radios that are not included in this book, it is not because they are undesirable, unattractive, or uncollectible. On the contrary, some of the sets not shown are among the most beautifully designed, colorful, and desirable radios of the collecting field. For example, the coveted Catalin models are among the rarest and most valuable of the plastic table radios, sometimes found in the $500 to $5,000 range.

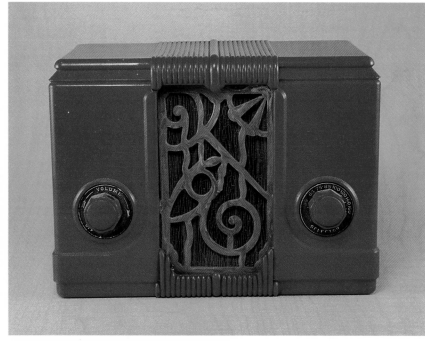

Example of Zenith early plastic radio with Clark Gable in promotional photo. *Courtesy Zenith Electronics Corporation.*

Example of early plastic radio. Kadette. 44. 1935. Red. *Courtesy Brett & Lori Gundlach.* **$150**

Example of novelty radio. Owl transistor radio (Natalie Jupp's favorite). c. early 1960s. White, gold. **$50**

Example of pocket transistor radio. Global. GR900. 1963. Red. $45

We have also chosen to omit very early plastic sets, and we have limited the number of basic brown and white painted Bakelite models. Although the lines may be great, a shelf filled with brown radios is not colorful enough. Other plastics not included are novelty radios -- radios in the form of people, animals, and objects such as toys or automobiles. That includes the advertising specialty, which usually takes the form of a consumer product and is often used as a premium or promotional give-away. Other novelty items are household furnishings, such as lamps, made from or to accommodate a radio.

An offshoot of radio collecting is radio-related items. Tube radios needed maintenance, and millions of replacement tubes were manufactured. Their boxes, with their interesting graphics, have become collectible in themselves. Signage and other items with early company logos are also of interest, but are not the subject of this book.

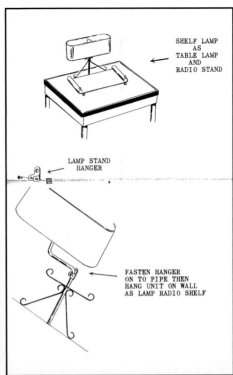

SHELF LAMP
AS
TABLE LAMP
AND
RADIO STAND

LAMP STAND
HANGER

FASTEN HANGER
ON TO PIPE THEN
HANG UNIT ON WALL
AS LAMP RADIO SHELF

Far left: Wall/table lamp designed for radio. **$35**

Left: Instructions for radio lamp.

RCA radio tubes in boxes.

Radio tubes.

9

Top left:
Example of wood table top radio.
Detrola. Bluebird. c. 1936. Wood. *Courtesy Bob and Darlene Jupp.*

Left:
Example of Zenith early floor model radio with Rita Hayworth in promotional photo. *Courtesy Zenith Electronics Corporation.*

Bottom Left:
Example of late model wood floor radios in Majestic International Corporation radio advertisement.

Finally we have not covered wooden radios, a completely separate category that has captivated radio enthusiasts for years. One advantage of wood is that damage to the case can often be repaired, and the item restored to its original beauty. The wood floor model was actually a piece of furniture, and the art of the cabinetmaker is apparent in models with exotic veneers and embroidered grille cloths.

What we do present in all their beauty are plastic radios. Most of the radios shown on the following pages are typical, decorative, collectible, and available. They can be found at specialized radio events -- shows, swap meetings, or auctions. They can also be purchased from generalized sources -- antique shops (especially those focusing on modernism), malls, shows, flea markets, used furniture stores, thrift shops, tag sales, or an old favorite -- the garage sale. Condition is usually good or excellent. Some have minor damage that is visible in the photographs, because our intent is to show these fascinating little objects as they are and as collectors are likely to find and enjoy them.

The values included at the end of each caption, however, are for radios of the same description in perfect condition and working order, so readers should bear in mind that flaws -- minor damage, re-

Right:
Example of late model solid state plastic radio. Merc-Radio. HT-601. c. late 1960s. Yellow, woodgrain. $10

Bottom right:
Zenith's Little Gold Box was introduced in 1939. This radio, Model 6D315, featured Zenith's Wavemagnet antenna for interference-free reception. Requiring no batteries or installation, it plugged into any AC power source to provide clear listening pleasure. *Courtesy Zenith Electronics Corporation.*

placements, refinishing, missing knobs, chips, cracks, or non-working condition -- will lessen value. That is not to say that only radios in perfect condition should be collected; we are saying that the values given are for perfect or almost perfect radios. In a collecting field, like this one, that is not well-established, some prices are apt to be higher or lower than those offered in this book. *Neither the authors nor the publisher can take any responsibility for any transactions based on consulting this guide.* Like any guide, it is primarily to help distinguish between the relatively common and rare pieces. These are prices that a collector might expect to pay for a similar radio. Selling is a different matter. Besides, by the time this book is printed, the market might have already changed. Monetary considerations aside, we wish you success and fun in your quest for genuine plastic radios.

Leslie Piña, Ph.D.

Ken Jupp

August, 1997

The Radios

Radios without Clocks

Zenith. 6-D-615. 1942. Brown Bakelite. **$100**

Zenith. 1938. Brown Bakelite. **$100**

Zenith. 6-D-312. 1938. Brown Bakelite. **$100**

Majestic. 250W. 1939. Brown Bakelite. **$75**

RCA. 65-X-1. 1948. Brown Bakelite. **$50**

Stromberg-Carlson. 1400. 1949. Brown Bakelite. **$100**

Airline. 1946. Brown Bakelite. **$75**

Fada. 605. 1946. Brown Bakelite. *Courtesy Brett & Lori Gundlach.* **$50**

Fada. 845. 1950. Brown Bakelite.
Courtesy Brett & Lori Gundlach. **$200**

Detail.

Opposite:
Delco. R 1151. c. 1940.
Brown Bakelite. *Courtesy
Brett & Lori Gundlach.*
$125

Stewart-Warner. 007-51H. 1940. Dark brown Bakelite. *Courtesy Brett & Lori Gundlach.* **$75**

General Electric. H510. 1939. Brown Beetle. *Courtesy Barry Gould.* **$150**

Aircastle. 10003. 1949. Brown Bakelite. *Courtesy Brett & Lori Gundlach.* **$100**

Gilfillan. 58M. 1948. Black marble. *Courtesy Barry Gould.* **$150**

Admiral. 7T10. 1947. Black. **$25**

Bendix. 526 A. 1946. Brown Bakelite. **$75**

Delco. R-1231. 1948. Brown Bakelite. *Courtesy Barry Gould.* **$75**

Silvertone. 7004. 1941. Brown. *Courtesy Barry Gould.* **$75**

Jewel. 955. 1950.
Black. *Courtesy
Barry Gould.* **$50**

Far left:
Emerson. 414. 1941.
Brown Bakelite. **$50**

Left:
Detail.

Emerson. 543. 1947.
Brown Bakelite. *Courtesy
Mark Klco.* **$50**

Right:
Emerson Radio and
Phonograph Corporation
magazine advertisement in
*The Saturday Evening
Post*, December 6, 1947.
Courtesy Emerson.

Far right:
Detail.

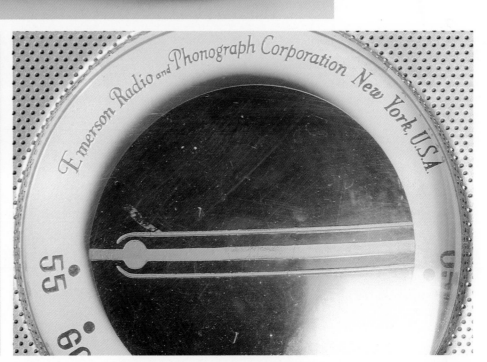

Top left:
Philco Transitone. 63-580. 1952. Black. **$25**

Top right:
Philco. 53-560. 1952. Ivory. **$25**

Bottom left:
Zenith. L721. 1954. Brown. **$25**

Top left:
Philco Transitone. 51-530-121.
1951. Brown Bakelite. **$25**

Top right:
Firestone. S-7402-4. 1940. Brown
Bakelite. **$50**

Right:
Detrola. 218 "Pee Wee". 1939.
Brown Bakelite, ivory details. **$300**

Above:
Silvertone. 9000. 1949.
Brown Bakelite. **$40**

Right:
Detail.

Belmont. 6D111. 1946. Painted ivory. **$150**

Sentinel. 329-I. 1947. Painted ivory. *Courtesy Barry Gould.* **$75**

Ambassador. c. late 1940s. Ivory. **$75**

Fada. 1947. Ivory. *Courtesy Barry Gould.* **$75**

Setchell-Carlson. 416. 1946. Ivory. **$75**

Opposite:
Stromberg-Carlson. 1101. 1946. Ivory.
Courtesy Barry Gould. **$50**

26

Crosley. 9-120W. 1948. White. *Courtesy Brett & Lori Gundlach.* **$75**

Crosley. 58 TK. 1948. Brown Bakelite. **$50**

Opposite top left:
Meck. 5E7-X12. 1948. Painted ivory.
Courtesy Barry Gould. **$75**

Opposite top right:
Mirror-Tone. 4B7. 1948. Painted ivory.
Courtesy Barry Gould. **$50**

Opposite bottom left:
Stewart-Warner. A51T3 "Air Pal". 1947.
Painted ivory. *Courtesy Brett & Lori
Gundlach.* **$100**

Opposite bottom right:
Trav-ler. 5000. 1946. Painted ivory.
Courtesy Brett & Lori Gundlach. **$50**

Bendix. 55L3. 1949. Ivory. *Courtesy Barry Gould.* **$75**

Truetone. D 1012. 1941. Ivory. *Courtesy Barry Gould.* **$75**

Westinghouse. H-743T4. 1951. Painted ivory. **$35**

Philco. 52-542. 1952. Painted white. *Courtesy Larry Jupp.* **$40**

Crosley. 9-104. 1949. Painted ivory, gold grille. **$50**

Clarion. Late 1940s. Ivory. **$50**

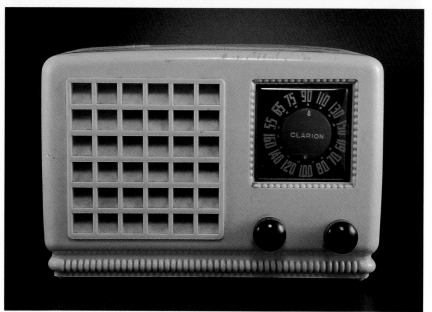

Olympic. 6A-501-U. 1946. White. *Courtesy Brett & Lori Gundlach.* **$75**

Garod. 7A-2. 1946. Ivory, blue grille.*Courtesy Barry Gould.* **$75**

Arvin. 651T. 1954. Painted ivory. **$50**

Arkay. 1948. White. *Courtesy Barry Gould.* **$75**

Above:
Motorola. 5R1. 1950.
Dark burgundy. *Courtesy
Barry Gould.* **$50**

Right:
Detail.

Opposite:
Dewald. H-410. 1954. White.
*Courtesy Brett & Lori
Gundlach.* **$75**

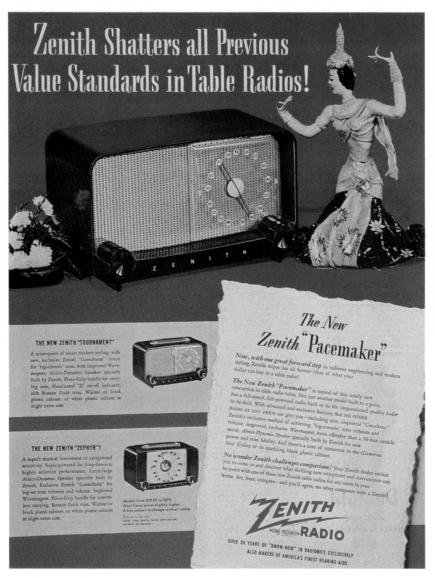

Above:
Zenith. G510. 1950. Black, gold grille. **$35**

Left:
Zenith Radio Corporation magazine advertisement
in *The Saturday Evening Post*, 1948. *Courtesy
Zenith Electronics Corp.*

Opposite:
Admiral. 5E22 AN.
1951. Brown, gold. **$50**

Emerson. 587A. 1949. Brown. **$50**

Sentinel. 338. 1951. Brown. *Courtesy Barry Gould.* **$35**

Truetone. D-2026. 1950. Brown. **$35**

Farnsworth. ET-060. 1946. Brown, ivory. **$50**

Silvertone. 3001. 1954. Brown. *Courtesy Barry Gould.* **$40**

Opposite:
Zenith. Y723. 1956.
Brown, ivory. **$50**

Right:
Detail.

Packard Bell. 5R1. 1957. Black. *Courtesy Barry Gould.* **$50**

General Electric. 440. 1954. Brown. **$25**

Westinghouse. H-310T5. 1950. Black. *Courtesy Barry Gould.* **$35**

Emerson. 572. 1949. Black. *Courtesy Barry Gould.* **$100**

RCA. 75X17. 1948. Gold chinoiserie. *Courtesy Barry Gould.* **$100**

Arvin. 850T. 1955. Brown. *Courtesy Barry Gould.* **$50**

Detail of chinoiserie from side of radio.

Detail.

41

Emerson. 520. 1946. Ivory Catalin. *Courtesy Dave Jupp.* **$100**

RCA. 9-X-571. 1950. Gold bullhorn, woodgrain. *Courtesy Barry Gould.* **$50**

Truetone. D2017. 1950. Brown Bakelite. *Courtesy Brett & Lori Gundlach.* **$125**

Left:
Detail.

Right:
Philco. 48-230 "Flying
Wedge". 1948. Chocolate
brown, ivory. **$100**

Right:
Philco. 49-503 "Flying Wedge".
1948. Black, ivory. *Courtesy*
Brett & Lori Gundlach. **$100**

Opposite:
Philco. 49-501 "Boomerang".
1949. Brown Bakelite. *Courtesy*
Brett & Lori Gundlach. **$225**

Crosley. 11-115 U. 1951. Dark red. *Courtesy Brett & Lori Gundlach.* **$125**

Crosley. 11-117 U. 1951. Dull green. **$125**

Top:
Crosley. 11-103 U. 1951. Red. *Courtesy Brett & Lori Gundlach.* **$150**

Above:
Detail.

Crosley Division, Avco Manufacturing
Corporation magazine advertisement, 1951.

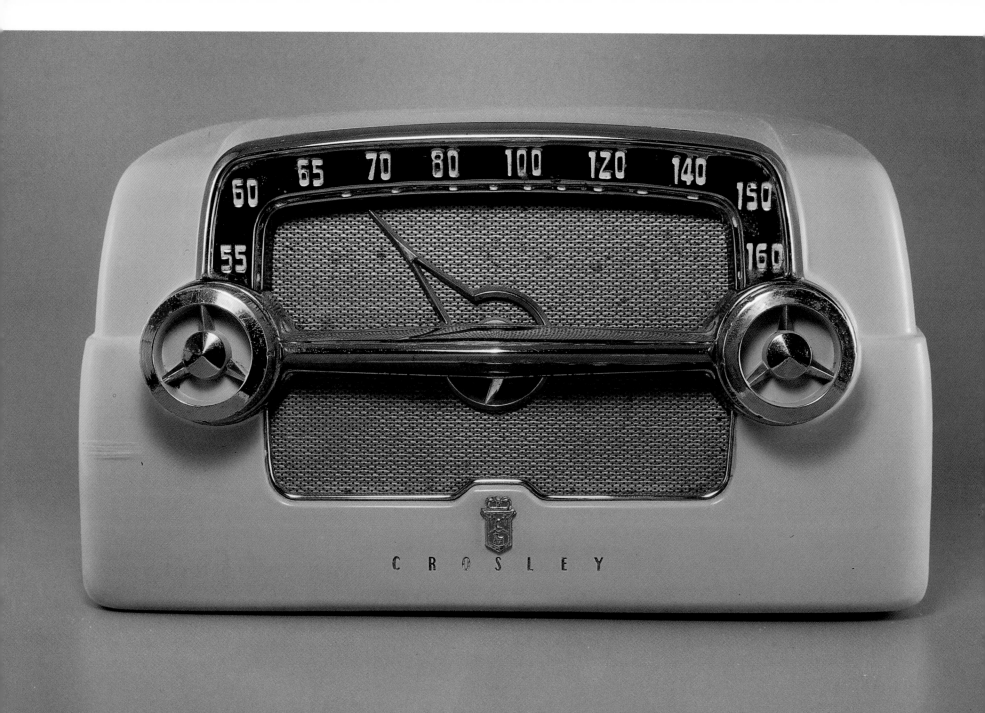

Above:
Crosley. 56TD. 1947. Burgundy, gold.
Courtesy Brett & Lori Gundlach. **$125**

Right:
Crosley. E-15TN. 1953. Tan. **$100**

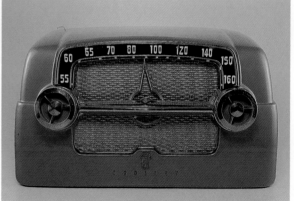

Opposite:
Crosley. E-15CE. 1953. Celery green.
Courtesy Barry Gould. **$100**

Crosley. 11-111 U. 1951. Olive green. *Courtesy Barry Gould.* **$125**

Crosley. 11-105 U. 1951. Light lime green. *Courtesy Barry Gould.* **$150**

Detail.

Crosley. 10-138. 1950. Silver gray. **$75**

Crosley. 10-137. 1950. Chartreuse. **$75**

Detail.

Left:
Crosley. 11-118 U. 1951. Taupe.
Courtesy Brett & Lori Gundlach. **$125**

Bottom left:
Crosley. 11-100 U. 1951. Ivory. *Courtesy
Brett & Lori Gundlach.* **$125**

Below:
Crosley. 10-135. 1950. White. **$75**

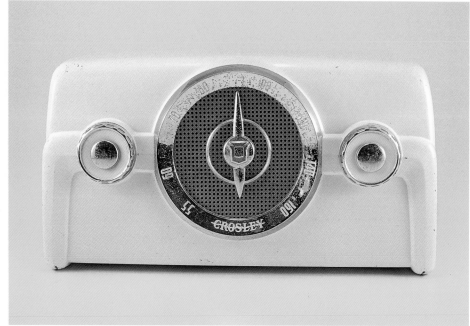

Above:
Crosley. 10-138. 1950.
Brown. **$75**

Top right:
Crosley. 11-102 U. 1951.
Metallic green. **$150**

Right:
Crosley. 11-108 U. 1952.
Brown. **$50**

Tele-Tone. 195. 1949. Maroon swirl. *Courtesy Barry Gould.* **$50**

Detail.

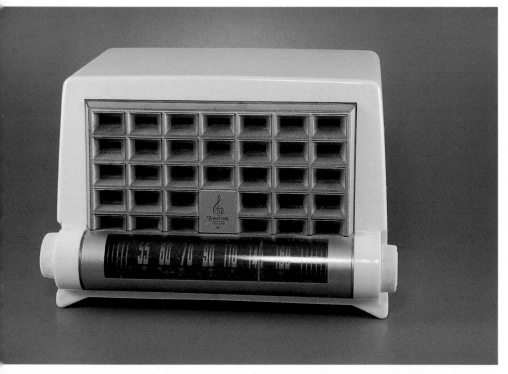

Arvin. 850T. 1956. Black. *Courtesy Barry Gould.* **$35**

Detail.

Opposite top left:
Emerson. 636A. 1950. Ivory, gold. **$50**

Opposite top right:
Clarion. 11802. 1947. Gray, burgundy.
Courtesy Brett & Lori Gundlach. **$50**

Opposite bottom left:
Arvin. 451-T. 1950. Khaki. *Courtesy
Barry Gould.* **$40**

Opposite bottom right:
Emerson. 547A. 1947. Butterscotch.
$50

Silvertone. 2014. 1956. Dark Green. **$25**

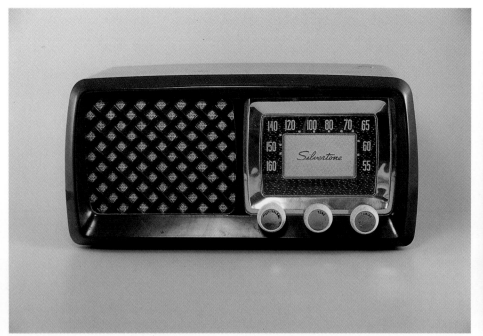

Detail.

Opposite:
Tele-Tone. 138. 1947. Brown,
butterscotch. **$75**

Trav-ler. T-202. 1959. Ivory, turquoise. *Courtesy Brett & Lori Gundlach.* **$50**

Trav-ler. T-201. 1959. Ivory, pink. *Courtesy Brett & Lori Gundlach.* **$50**

Opposite top left:
Trav-ler. T-200. 1959. Black, pink. **$50**

Opposite top right:
Trav-ler. T-200. 1959. Black. **$50**

Opposite bottom lerft:
Detail.

Opposite bottom right:
Detail.

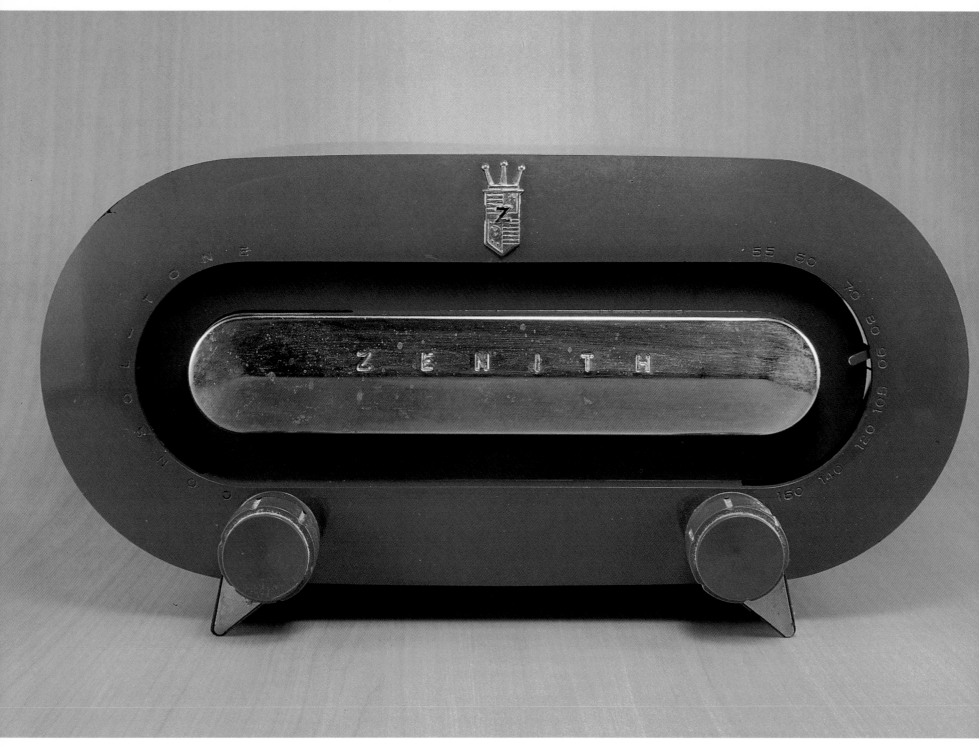

Zenith. H511. 1951. Green. *Courtesy Brett & Lori Gundlach.* **$75**

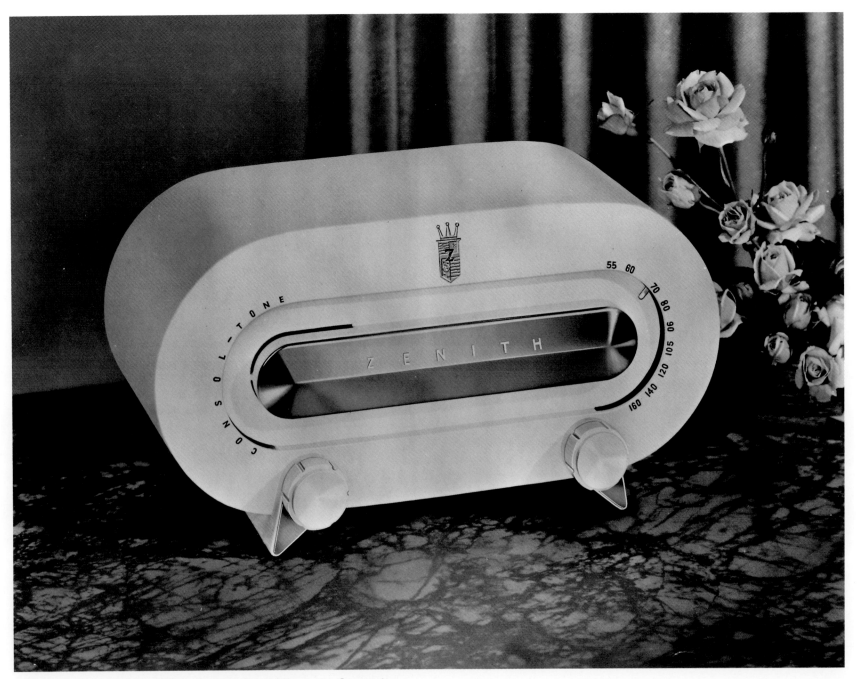

Zenith promotional photo of H511. *Courtesy Zenith Electronics Corporation.*

Philco. 53-561. 1952. Putty. *Courtesy Barry Gould.* **$35**

Sterling Deluxe. Late 1940s. Blue swirl, black. **$75**

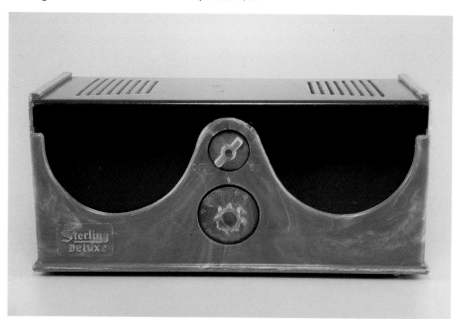

Sylvania. 568. 1955. Dark green. **$30**

Philco. 53-566. 1952. Maroon. *Courtesy Barry Gould.* **$75**

Arvin. 741T. 1953. Red. *Courtesy Barry Gould.* **$40**

Coronado. 43-8245. c. 1950s. Red. *Courtesy Barry Gould.* **$50**

Philco. B570. 1954. Red. *Courtesy Barry Gould.* **$40**

RCA. 6-X-8B. 1955. Red, gold. *Courtesy Barry Gould.* **$75**

Crosley. E-15CE. 1953. Celery green. *Courtesy Barry Gould.* **$75**

Detail.

Olympic. 442W-1. c. late 1950s. Light blue. *Courtesy Barry Gould.* **$50**

Detail.

Above:
Detail.

Right:
Arvin. 840T. 1995. Red
painted metal, taupe
plastic details. **$45**

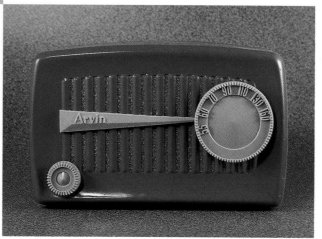

Above:
Motorola Radio magazine advertisement in *The Saturday Evening Post*, 1953. *Courtesy Motorola Inc.*

Left:
Motorola. 53H. 1954. Red-orange. *Courtesy Barry Gould.* **$75**

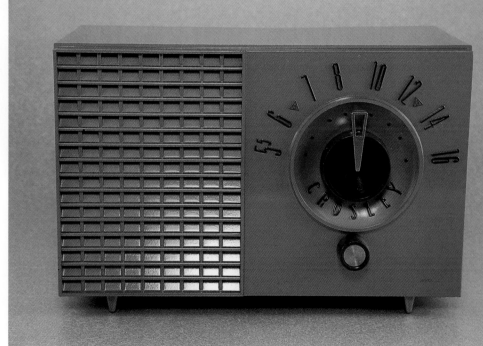

Top left:
Crosley. F-5CE. 1959.
Chartreuse. **$50**

Top right:
Crosley. F-5TWE "Musical
Chef". 1959. White.
Courtesy Barry Gould. **$50**

Right:
Crosley. 1954. Metallic
copper. **$50**

Opposite:
Crosley. F-5RD.
1959. Red. **$50**

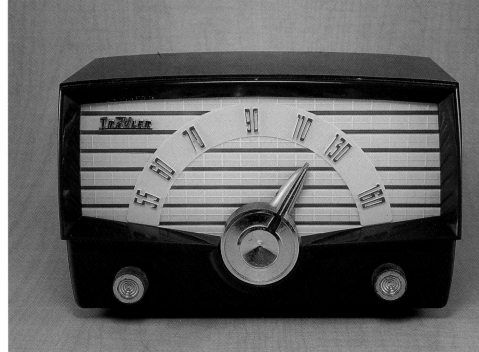

Capehart. T-522. 1953.
Blue, ivory. *Courtesy
Barry Gould.* **$40**

Opposite top left:
Zenith. A513R. 1957. Ivory, dark
burgundy. *Courtesy Brett & Lori
Gundlach.* **$35**

Opposite top right:
Zenith. Y513F. 1956. Ivory, light green.
$35

Opposite bottom left:
Trav-ler. 55-38. 1955. Yellow, Green.
Courtesy Brett & Lori Gundlach. **$50**

Opposite bottom right:
Trav-ler. 55-38. 1955. Ivory, burgundy.
Courtesy Barry Gould. **$40**

Detail.

Philco. B570. 1954. Stop red. *Courtesy Barry Gould.* **$50**

Sentinal. 352. 1956. Red-orange. *Courtesy Barry Gould.* **$45**

Arvin. 2581. 1958. Black, White. **$35**

Crosley. T-41BK. 1956. Black, silver. **$35**

Airline. GSE 1622A. 1956. Caramel swirl. *Courtesy Barry Gould.* **$50**

Left: Emerson. 706. 1952. Gold, ivory. **$50**
Right: Emerson. 707. 1952. Gold, ivory. *Courtesy Barry Gould.* **$75**

Detail.

Detail.

Top: Westinghouse. H-381T5. 1953. Red.
Bottom: Westinghouse. H-380T5. 1953. Green.
Courtesy Barry Gould. **$30 each**

84

Top: Emerson. 708. 1952. Red.
Bottom: Emerson. 708. 1952. Black.
Courtesy Barry Gould. **$35 each**

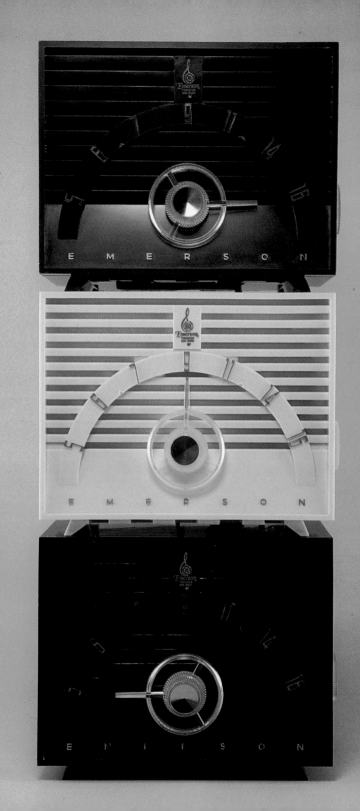

Above:
Emerson. 708. 1952.
Light pink. *Courtesy Barry Gould.* **$35**

Right:
Top: Emerson. 811. 1955.
Spruce green.
Center: White. Bottom:
Black. *Courtesy Barry Gould.* **$35 each**

Top: Motorola. 5T21W-1. 1957. White.
Bottom: Motorola. 57-A. 1957. Black. *Courtesy Barry Gould.* **$30 each**

Top: Motorola. 56-R. 1956. White.
Bottom: Motorola. 56-R. 1956. Red. *Courtesy Barry Gould.* **$40 each**

Motorola. 5-T-22 "Dragster". 1957. *Courtesy Barry Gould.* **$75**

Zenith. Z512G. 1958. Gray. *Courtesy Barry Gould.* **$35**

Zenith. Y511R. 1956. Burgundy. **$35**

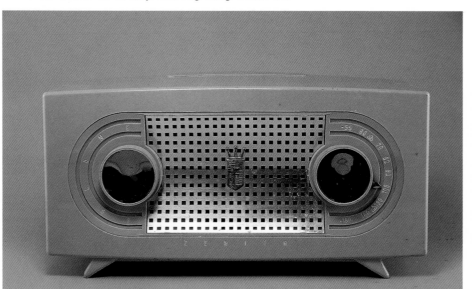

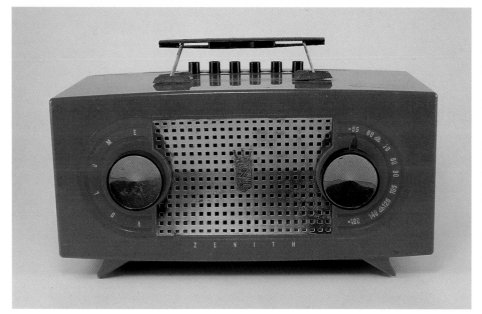

Zenith. R512V. 1955. Red. **$40**

Detail.

General Electric. T125A. 1958. Pink. **$25**

Stewart. c. 1950s. Pink. *Courtesy Barry Gould*. **$25**

Dumont. R1130-8962. c. late 1950s. Pink, gold. **$30**

Zenith. Z510G. 1957. Gray. *Courtesy Barry Gould*. **$45**

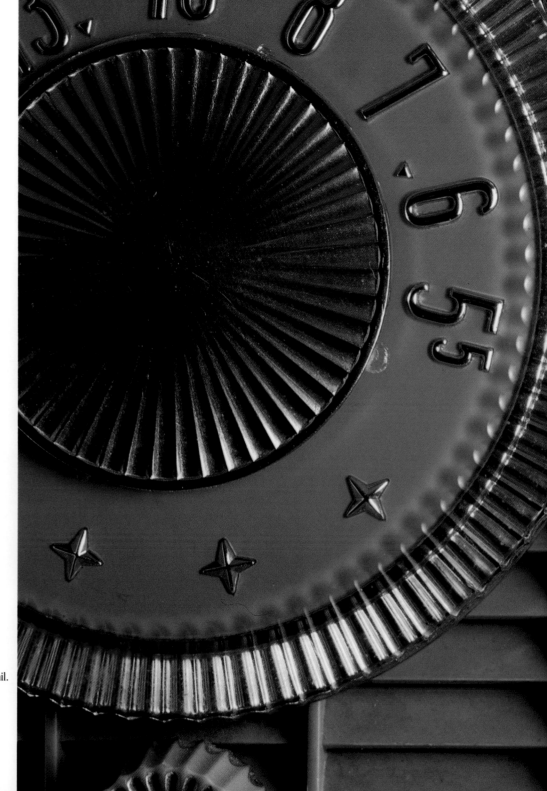

Opposite:
Bulova. 320. c. early
1960s. Turquoise.
Courtesy Barry Gould.
$25

Detail.

Philco. E-814124. 1957. Green. **$30**

Detail.

Detail.

Motorola. 57R2. 1956. Ivory. *Courtesy Barry Gould.* **$40**

Philco. C-587. 1955. Salmon. **$30**

Motorola. 52R. 1952. Yellow (paint not original). *Courtesy Barry Gould.* **$35**

Admiral. Y-2998. 1961. Blue. **$20**

RCA. 8-X-8L. 1957. Light blue. **$25**

Emerson. 652-B. 1950. Ivory. **$25**

Airline. GSE 1526A. c. late 1950s. Turquoise. *Courtesy Barry Gould.* **$30**

Detail.

Top left:
Westinghouse. H-742T4.
1960. Turquoise. *Courtesy Barry Gould.* **$25**

Top right:
Westinghouse. H-576T4.
1956. Pink. *Courtesy Barry Gould.* **$25**

Right:
Westinghouse. H-743T4.
1960. White. **$20**

Left:
Sylvania. 5151. 1956. Robin's-egg
blue. *Courtesy Barry Gould.* **$35**

Bottom left:
Arvin. 951T. 1955. Red. *Courtesy
Barry Gould.* **$30**

Bottom right:
Meteor. 7000. 1957. Slate gray.
Courtesy Barry Gould. **$20**

Motorola. 56H. 1956. Mint green. *Courtesy Barry Gould.* **$75**

Admiral. 5G35N. 1951. Red. *Courtesy Brett & Lori Gundlach.* **$30**

Admiral. 5G31. 1951. Black. *Courtesy Brett & Lori Gundlach.* **$25**

Motorola. 56H. 1956. Olive green. *Courtesy Barry Gould.* **$75**

Above:
Emerson. 915. 1960. Sky blue.
Courtesy Barry Gould. **$30**

Right:
Detail.

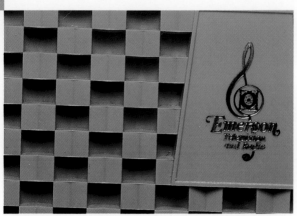

Emerson. 778. 1954. Red. *Courtesy Barry Gould.* **$50**

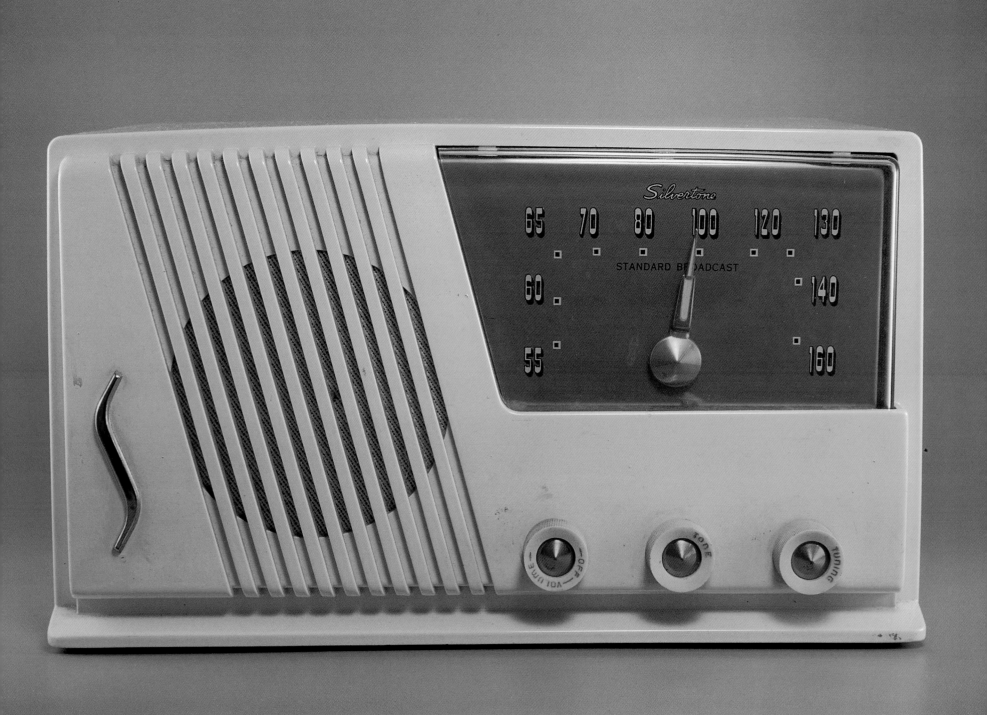

Silvertone. 16. 1951. White. *Courtesy Barry Gould.* **$35**

Admiral. Y-3318. 1963. Turquoise. *Courtesy Barry Gould.* **$20**

Detail.

Opposite top left:
Arvin. 480-TFM. 1950. Taupe.
$25

Opposite bottom left:
Detail.

Opposite top right:
Firestone. 4-A-159. 1957.
Two-tone green. **$30**

Opposite bottom right:
Detail.

Motorola. 56-R. 1956. Ivory. **$40**

Detail.

Detail.

Motorola. A-1 R2. 1960. Red. *Courtesy Barry Gould.* **$25**

Motorola. A15W. 1962. White. **$15**

Motorola. A16P. 1962. Pink, white. **$20**

Motorola. A-8. 1961. Turquoise. *Courtesy Barry Gould.* **$25**

General Electric. 429. 1955. Red. **$35**

Trancel. UL-518. c. early 1960s. White, burgundy. **$25**

Western Auto (Truetone). DC-2630A. c. late 1950s.
Mint green, ivory *Courtesy Barry Gould*. **$25**

Sylvania. 519. 1956. Turquoise, white. *Courtesy Barry Gould*. **$30**

Silvertone. c. early 1960s. Red, white. **$15**

Philco. G826-124. 1959 . Turquoise.
Courtesy Barry Gould. **$30**

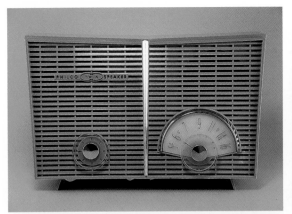

Westinghouse. H-648TA. 1959. White. **$20**

Westinghouse. H-438T5. 1958. Light green. **$25**

Zenith. B508R. 1957. Dark burgundy, ivory. **$25**

Emerson. 812. 1955. Gray. *Courtesy Bill Igo.* **$25**

Motorola. AT31BN. c. mid-1960s.
Courtesy Barry Gould. **$20**

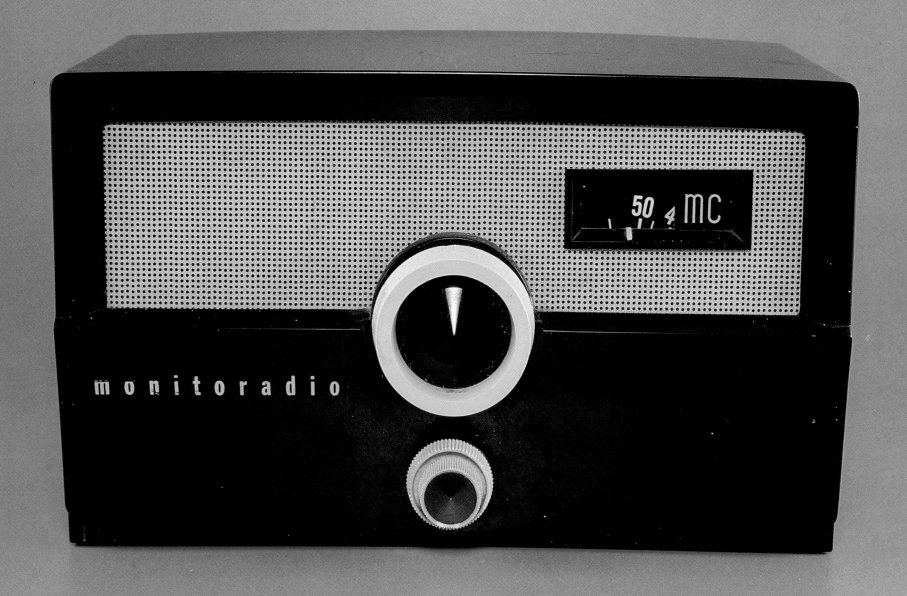

Monitoradio. (Police monitor). c. 1950s. **$20**

Above:
Emerson. 870. 1958. Pink, black. **$25**

Left:
Emerson. 851. 1957. Black. **$20**

General Electric. 860. 1957. Red, white,
black. *Courtesy Barry Gould.* **$50**

Detail.

Top left:
Alco. c. early 1960s. Blue,
chrome. *Courtesy Brett &
Lori Gundlach*. **$30**

Top center:
Omscolite. c. early 1960s.
Pink, gold. **$30**

Top right:
Monarch. c. early 1960s.
Black, white. *Courtesy Brett
& Lori Gundlach*. **$25**

Right:
Detail.

Zephyr. c. early 1960s. White. **$20**

Classic. c. mid-1960s. Yellow. *Courtesy Barry Gould.* **$20**

Golden Bell. c. 1960s. Pink. *Courtesy Brett & Lori Gundlach.* **$20**

Galaxie. c. late 1960s. Red. **$20**

Sylvania. 1108. 1959. Deep turquoise, white. *Courtesy Barry Gould.* **$25**

Major. c. early 1960s. White, turquoise. *Courtesy Bill Igo.* **$25**

Right:
Channel Master. 6534. c. early
1960s. Dark teal, white. **$25**

Below left and right:
Details.

117

Westinghouse. H-637T6A. 1958. Coral, white. *Courtesy Jim Vatter.* **$15**

Detail.

Channel Master. 6511. c. early 1960s. Light Green. **$20**

Channel Master Radios key chain. **$2**

Silvertone. 8002. c. late 1960s. Putty. **$10**

RCA. 8-BX-5F. c. early 1970s. White. **$15**

Detail.

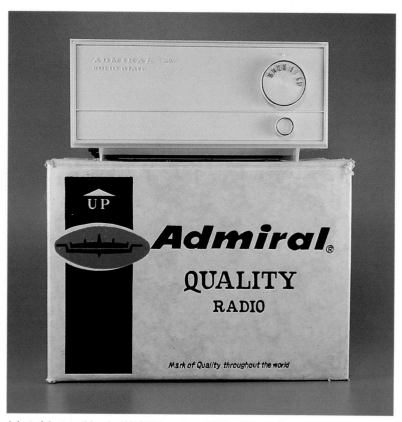

Raleigh Electronic Corp.. T-607. c. late 1960s.
Ocher. *Courtesy Brett & Lori Gundlach.* **$20**

Admiral (original box). AY107RA. c. late 1960s. White. **$10**

Opposite:
Merc-Radio (Original Box).
HT-601. c. late 1960s.
Yellow, woodgrain. **$10**

Clock radios

General Electric. c. 1940s. Brown Bakelite. **$25**

Westinghouse. H-374T5. 1952. Dark maroon. *Courtesy Barry Gould.* **$40**

Philharmonic. 51C3A. c. Late 1940s. Brown Bakelite. **$40**

Jewel. 920A. 1948. Brown Bakelite. **$25**

Zenith. G516Y. 1950. Black with gold trim. *Courtesy Bill Igo.* **$40**

Above:
General Electric. 515.
1951. Tortoise shell. **$35**

Right:
General Electric. 515F.
1951. Ivory. **$35**

General Electric Company, Receiver Division magazine advertisement in *The Saturday Evening Post*, 1948. *Courtesy General Electric.*

General Electric Company magazine advertisement in *Life Magazine*, April 14, 1947. *Courtesy General Electric.*

Sylvania. 540M. 1951. Brown, ivory. **$35**

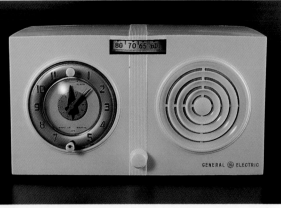

Above:
General Electric. 511. 1951.
Ivory. *Courtesy Bill Igo.* **$50**

Left:
Detail.

Sparton. 132. 1950. Maroon. *Not* a clock radio. Notice the similarity
to the Sparton "Eye Opener". *Courtesy Barry Gould.* **$75**

Sparton. "Eye-Opener". 1950. Ivory. **$75**

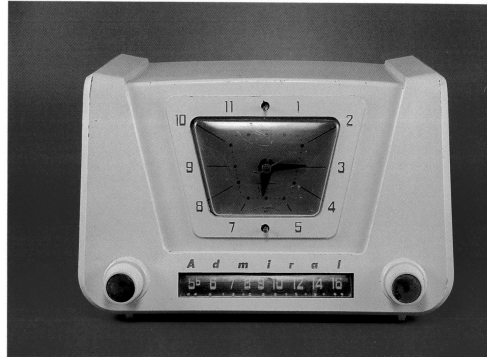

Top left:
General Electric. 549. 1953.
Brown. **$30**

Top right:
General Electric. 551. 1953.
Dark burgundy swirl. **$30**

Right:
General Electric. 573. 1955.
Ivory. **$30**

Opposite top left:
Stewart Warner. 9164. 1952. Gray, ivory. **$40**

Opposite top right:
Cavalier. 5C1. 1954. Green, gold. **$35**

Opposite bottom left:
Stewart Warner. 9162. 1952. Pale green. **$40**

Opposite bottom right:
Admiral. 5X23. 1953. Ivory. **$35**

CROSLEY

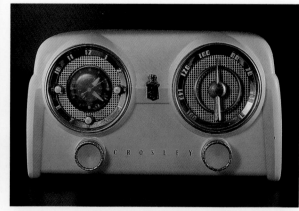

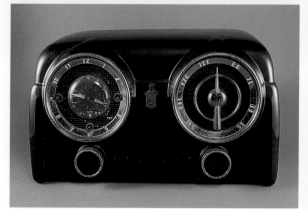

Above:
Detail.

Above right:
Crosley. D-25CE. 1953.
Chartreuse. *Courtesy Brett*
& Lori Gundlach. **$125**

Right:
Crosley. D-25BE. 1953.
Dark blue. **$125**

Opposite:
Crosley. D-25BE. 1953. Silver gray.
Courtesy Brett & Lori Gundlach. **$125**

Above:
Detail.

Right:
General Electric. 590.
1955. Brown, gold. **$30**

136

General Electric. 574. 1955. Red. **$35**

Zenith. J733. 1952. Brown, ivory. **$40**

Opposite:
Zenith. R514V. Coral.
1953. **$125**

Detail.

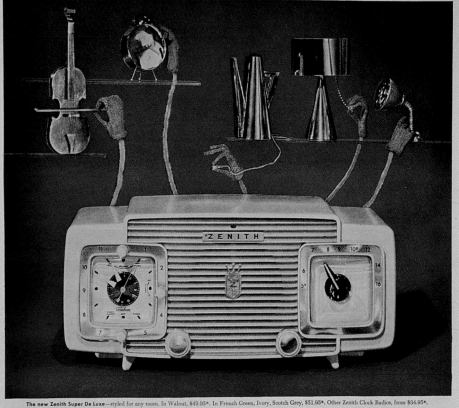

Above:
Zenith. K622. 1953. Green, gold. **$50**

Left:
Zenith Radio Corporation magazine advertisement in *The Saturday Evening Post,* March 12, 1953. *Courtesy Zenith Electronics Corp.*

140

Top left:
Silvertone. 8. 1951. Dark red.
Courtesy Barry Gould. **$35**

Top right:
Tele-tone. c. 1950s. Turquoise.
$50

Right:
Emerson. 778. 1954. Red.
Courtesy Barry Gould. **$50**

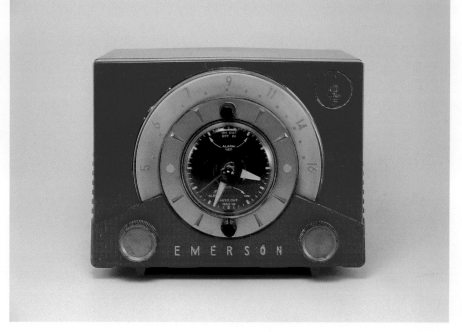

Motorola. 66C. 1956. Slate Gray. *Courtesy Barry Gould.* **$50**

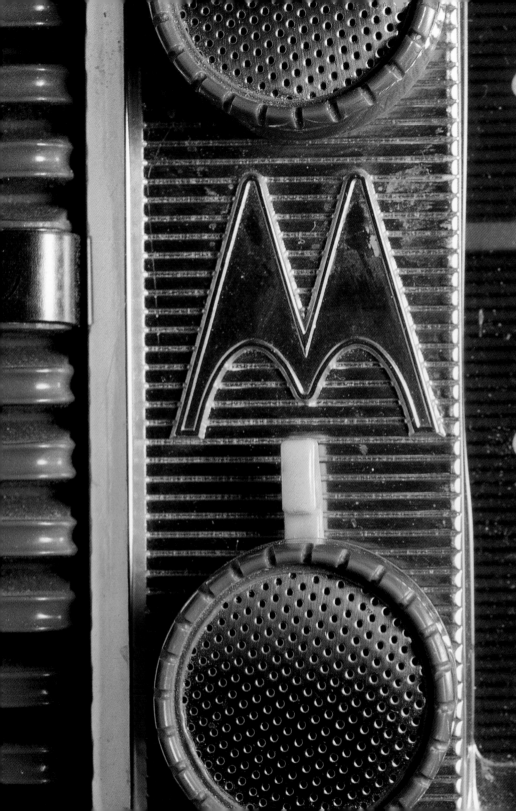

Left:
Detail.

Top right:
Detail.

Above:
Westinghouse. H-538T4A. 1955. Black. **$30**

143

Trav-ler. 55C46. 1957. Brown, gold. **$35**

Bulova. 100. 1961. Black, gold. **$45**

Crosley. E-85GN. 1953. Black, mint green. **$40**

Detail.

Capehart. 75C56. 1956. Red, gold. **$45**

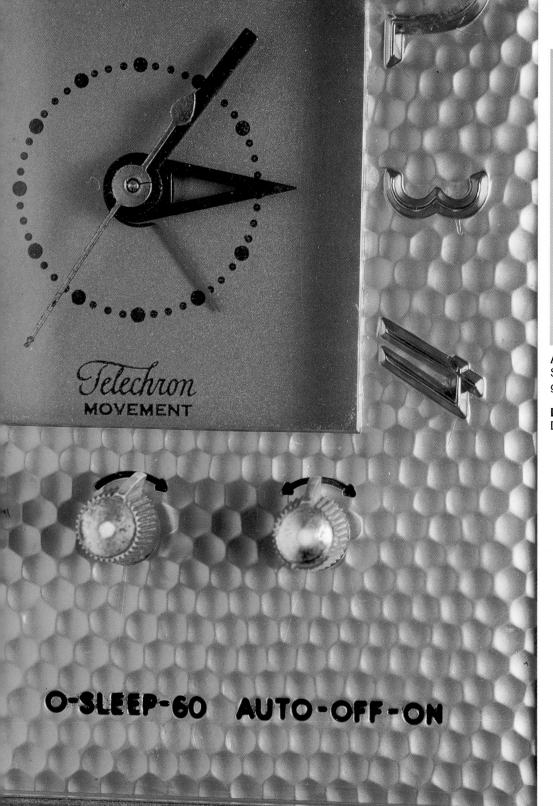

Telechron
MOVEMENT

0-SLEEP-60   AUTO-OFF-ON

Above:
Stromberg-Carlson. C-5. 1955. Mantis green,
gold. *Courtesy Barry Gould*. **$45**

Left:
Detail.

Opposite top left:
Emerson. 825-B. 1955. Red.
Courtesy Barry Gould. **$75**

Opposite top right:
Motorola. 57-CS. 1957. Mint
green. **$50**

Opposite bottom left:
Emerson. 826-B. 1955.
White, gold. *Courtesy John
Wichman*. **$35**

Opposite bottom right:
Firestone. 4-A-12?. 1954.
Coral orange. **$40**

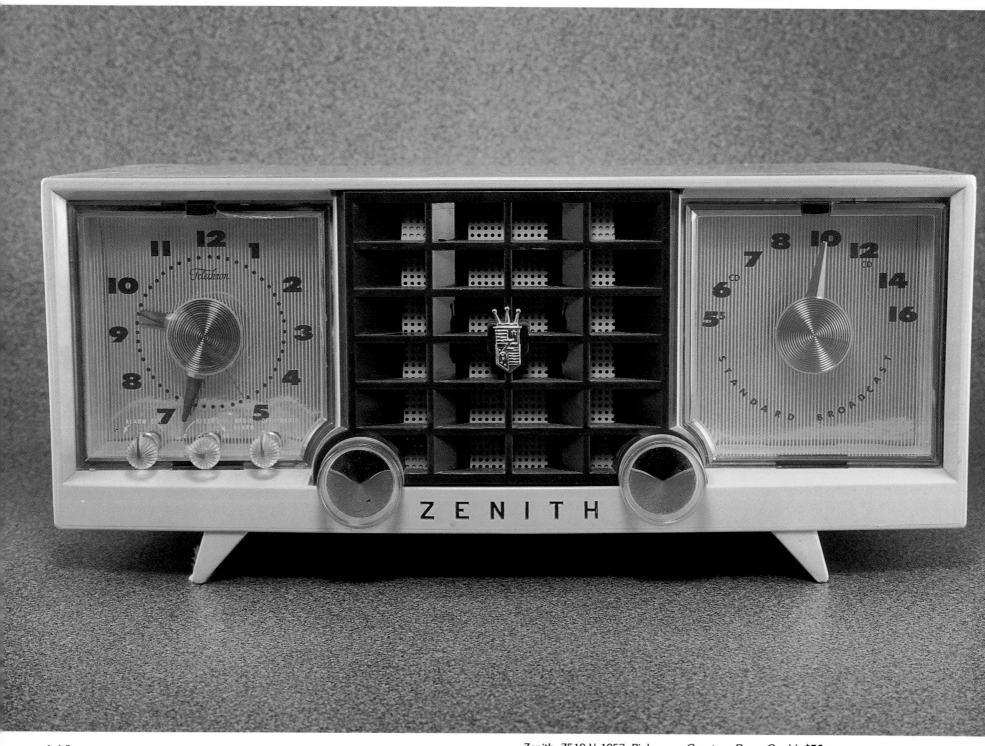

148

Zenith. Z519-V. 1957. Pink, gray. *Courtesy Barry Gould.* **$50**

Detail.

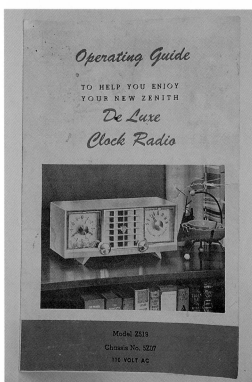

Zenith Operating Guide.

Zenith magazine advertisement in *The Saturday Evening Post*, October 8, 1955. *Courtesy Zenith Electronics Corp.*

Zenith ad slick.

Howard Pierce porcelain ducks shown in Zenith ad slick.

Opposite: Bulova. 170. 1961. Turquoise, gold. *Courtesy Barry Gould.* **$30**

150

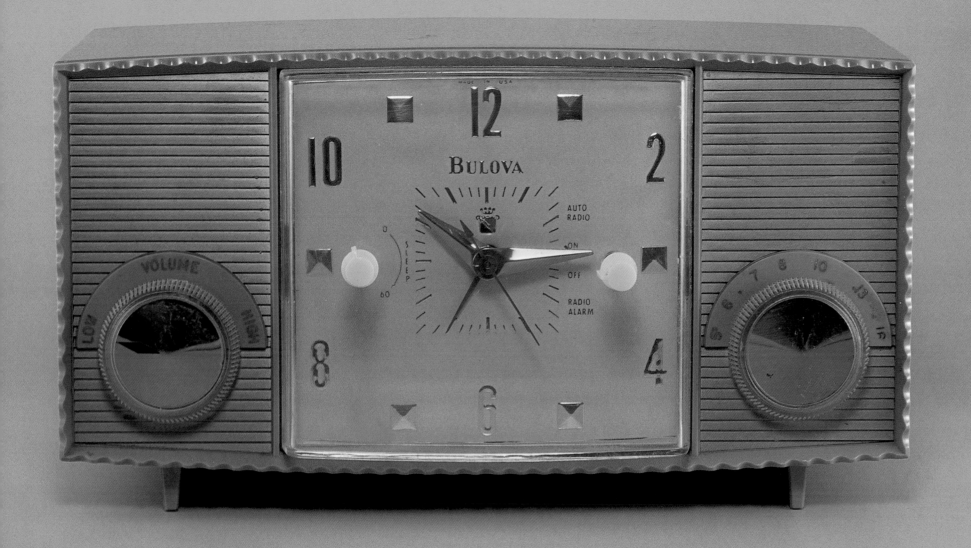

General Electric. 556. 1954. Red. **$50**

Detail.

Detail.

153

General Electric. C431A. 1958. Turquoise, black. **$35**

General Electric. C43?. 1958. Pink, black. **$25**

General Electric. 911D. 1957. Dark brown. **$25**

General Electric. C43. 1958. White, black. **$25**

Firestone. 4-A-160. 1956. Pink, black. *Courtesy Barry Gould.* **$35**

Emerson. G-1704. 1962. Black. *Courtesy Steve "Ba-Did" Dotts.* **$20**

Sylvania. 2107. 1959. Lemon yellow. **$35**

Detail.

Left:
General Electric.
C2418A. c. 1960. White,
Mickey Mouse. *Courtesy
Barry Gould.* **$100**

Above:
Detail.

Opposite
General Electric. C249A.
c. 1960. Chiffon,
Disneyland. *Courtesy
Barry Gould.* **$125**

Detail.

Top left:
Detail.

Top right:
General Electric. C-404B.
1957. Baby blue. *Courtesy
Barry Gould.* **$20**

Right:
General Electric. C-403G.
1961. White. **$15**

Detail.

Silvertone. 5035. 1965. Light pink. *Courtesy Barry Gould.* **$35**

Left:
Silvertone. 2079. c. 1962. Black,
gold, white. *Courtesy Bill Igo.*
$25

Bottom left:
Zenith. H519L. 1962. White,
fleshtone. **$15**

Bottom right:
Zenith Radio Corporation
magazine advertisement.

Silvertone. 2061. c. late 1960s. Off white. **$10**

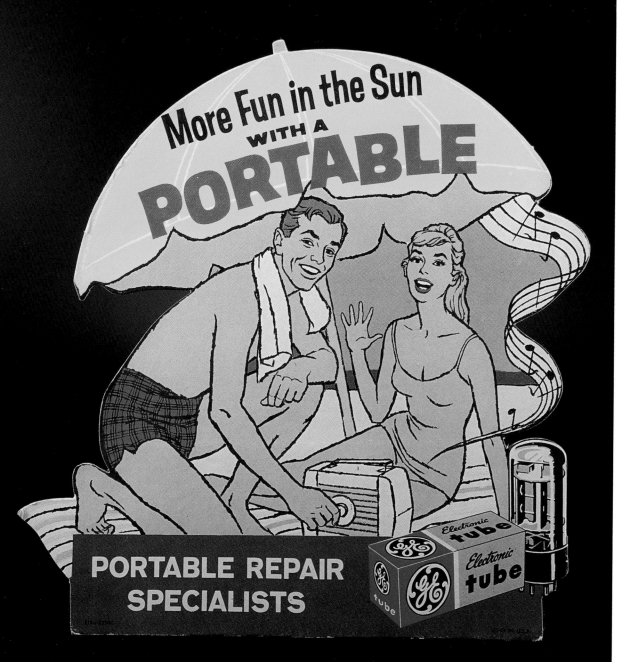

General Electric radio tube
point-of-purchase cardboard
display, c. 1950s.

General Electric radio tubes in boxes, c. 1950s.

Opposite:
Detail.

Right:
Zenith Radio Corporation
magazine advertisement in
The Saturday Evening Post,
February 7, 1953. *Courtesy
Zenith Electronics Corp.*

Below:
Zenith. K412R. 1953.
Brown, silver. $75

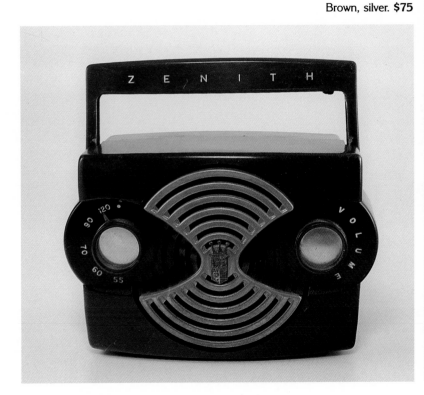

Zenith. J-402R. 1952. Burgundy, gold. *Courtesy Bill Igo.* **$50**

Top view.

Above:
Motorola. 48L11. 1948.
Burgundy, taupe, right knob
not original. *Courtesy Paul
"the Dugue" Oravetz.* **$40**

Left:
Motorola Radio magazine
advertisement, 1948.

Opposite:
Detail.

Motorola Radio magazine advertisement, 1951.

Motorola Radio magazine advertisement in *The Saturday Evening Post*, May 17, 1952. *Courtesy Motorola Inc.*

Opposite:
Motorola. 59L12Q. 1949.
Green, gold. **$40**

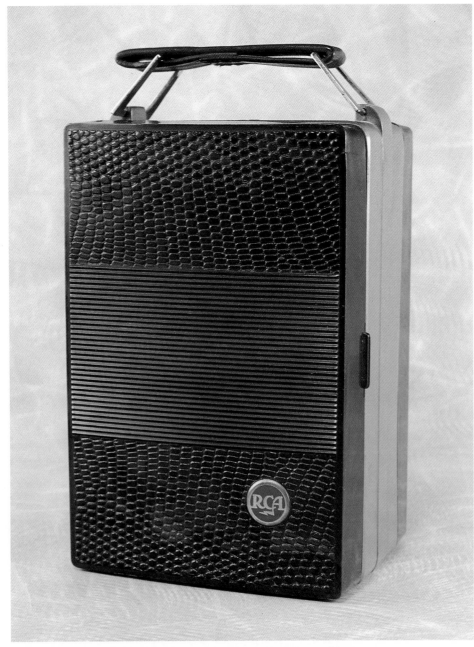

RCA. 8-B-42. 1949. Brown. Case closed. *Courtesy Dave Jupp.* **$50**

RCA Victor, Division of Radio Corporation of America,
magazine advertisement, 1947. *Courtesy RCA.*

Opposite:
case open.

Trav-ler. 5022. 1950. Ivory front, alligator case. **$40**

Back view, case open.

Sylvania. 433B. 1953. Green, gold. **$40**

Top view.

Arvin. 240P. 1948. Dark red. **$50**

Emerson. 559A. 1948. Dark Red. **$25**

RCA. BX-57. 1950. Alligator finish. **$25**

Emerson. 646A. 1950. Terra cotta. *Courtesy Barry Gould.* **$40**

Detail.

Detail.

Crosley Division, Avco Manufacturing Corporation
magazine advertisement, c. 1950s.

Trav-ler. 5300. 1953. Red, gold. **$35**

Zenith. T404F. 1955. Red, black. *Courtesy Darlene Jupp.* **$35**

Opposite:
Crosley. F100RD.
"Skymaster". 1953.
Red. *Courtesy Jack &*
Sylvia Schenkel. **$40**

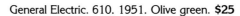
General Electric. 610. 1951. Olive green. **$25**

Detail.

RCA. 7-BX-8L. 1957. Dull teal, silver,
Courtesy Rae Cavanaugh. **$25**

Opposite:
Philco. 53-652. 1954.
Red. *Courtesy Barry
Gould.* **$40**

Zenith. K401. 1954. Black, gold, with leather carrying
case. *Courtesy Brett & Lori Gundlach.* **$50**

Sylvania. 454. 1954. Gray, with leather carrying case. **$40**

Arvin. 747-P. 1953. Gold, green. *Courtesy Brett & Lori Gundlach*. **$45**

Hallicrafters. TW-24. 1953. Green front, plaid case. **$35**

Detail.

Zenith. Royal 750L. 1959. Gold, leather case. **$10**

RCA. 8-BX-5F. 1948. Flesh-tone. **$15**

General Electric. P-761A. 1958. White, green. **$15**

RCA. 7-BX-6. 1956. Gray. **$15**

Crosley. P-50WE. c. late 1950s. White, red dots. *Courtesy Tim Poling.* **$25**

Transistors

Ad slick for Zenith Royal 500E Y transistor. *Courtesy Zenith Electronics Corporation.*

Zenith. Royal 500. 1955. Black, gold. **$50**

Top left:
Motorola "Pixie". 45P5. c. late
1950s. Turquoise. *Courtesy
Brett & Lori Gundlach.* **$100**

Top right:
RCA. c. early 1970s. Olive.
Courtesy Barry Gould. **$10**

Right:
Nipco. BL-006P. c. late 1950s.
Red. **$15**

Logos

Admiral

Admiral

Admiral

Airline

Alco

Ambassador

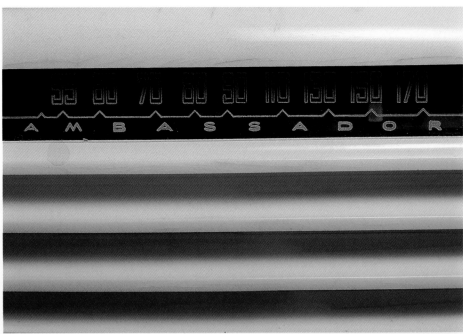

Arvin

Arvin

Bendix

Bulova

Capehart

Channel Master

Crosley

Crosley

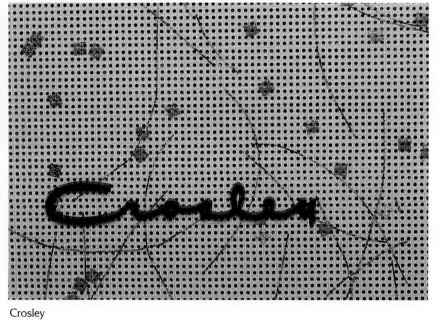

Crosley

Crosley

Crosley

Crosley

Crosley

Detrola

Emerson

Dumont

Emerson

Emerson

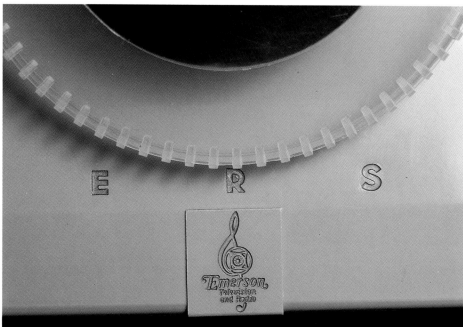

Fada

Fada

Firestone

Galaxie

Garod

General Electric

General Electric

Golden Bell

Hallicrafters

Jewel

Majestic

Merc-Radio

Monarch

Motorola

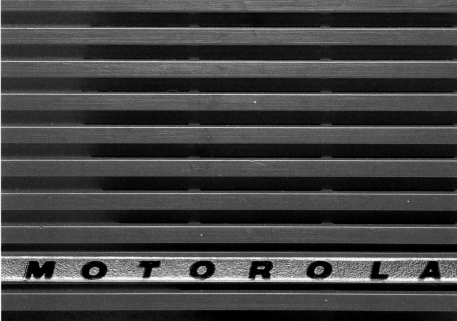

Motorola

Musaphonic

200

Omscolite

Philco

Philco Transitone

Philco "Twin Speaker"

Pixie

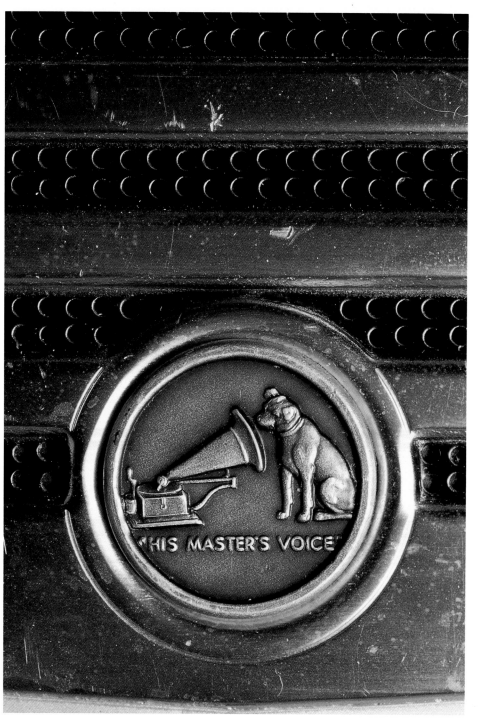

RCA "His Master's Voice"

Sears

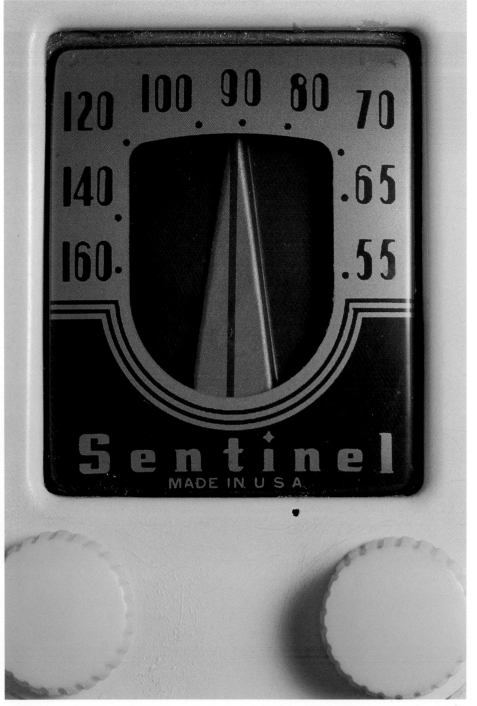

Sentinel

Silvertone

Silvertone

Sparton

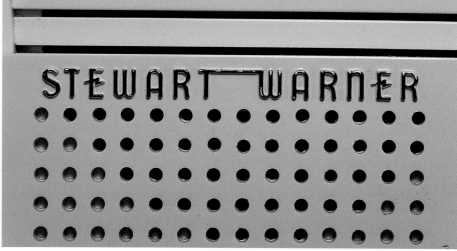

Stewart Warner

Sylvania

Sylvania

Sylvania

Symphonaire

Tele-tone

Tele-tone

55 60 70 80 100 120 140 170

Trav-ler

160 140 110 90 75 60 55

Trav-ler

Trav-ler

Truetone

Tune-riser

Westinghouse

Westinghouse

208

Zenith

Zenith

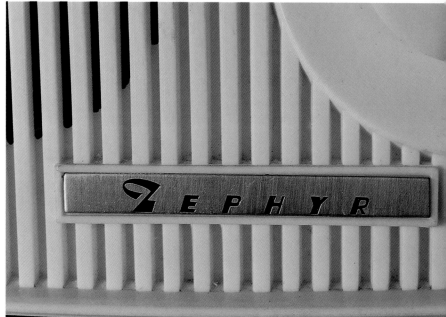

Above:
Zephyr

Left:
Zenith

Motorola Highlights

We had to learn all the hard way. We had no sponsorship, no money, no business. The car radio idea in 1930 was very unpopular...[yet] our business attracted young men who hoped to grow up with a new industry and were willing to take a chance.

Paul V. Galvin

On September 25, 1928, Paul V. Galvin (1895-1959) and his brother Joseph E. Galvin (1899-1944) incorporated the Galvin Manufacturing Corporation. Its first product, a battery eliminator, enabled battery-operated home radios to operate on ordinary household current. In 1930 the company produced the first practical and affordable automobile radio. The name Motorola was coined, linking the concepts of motion, automobiles, and radios.

In 1936, the AM automobile radio *Police Cruiser* was Motorola's first entry into the new field of radio communication products. In the following year, 1937, a new line of Motorola home radios was announced. The first commercial line of two-way FM radio communication products was introduced in 1941. Galvin Manufacturing Corp. officially changed its name to Motorola, Inc. in 1947, and in 1948 the company supplied radios for the big three automobile makers -- Chrysler, Ford, and General Motors.

At mid-century, Motorola was a major supplier of Bakelite, followed by colorful plastic table radios. After experimenting with transistors, in 1959 an all-transistor automobile radio was produced. In the same year, the company introduced a shirt-pocket-size all-transistor portable radio. The last Motorola automobile radio was made in 1987, and today the company continues to develop and manufacture electronics for consumer markets and communications as well as for government and space exploration. The Motorola Museum of Electronics opened in Schaumburg, Illinois in 1991.

1930

1941

1947

1955

1967

Zenith Highlights

The history of Zenith is to a considerable degree a history of the radio-television industry. This is so because Zenith has been a pioneer and leader in radionics since before there was a radio industry, and has played an important role in almost every important development during radio growth from an amateur toy to the most significant, widespread, and effective system of communications in history.

The Zenith Story 1955

In 1918, amateur radio operators R.H.G. Mathews and Karl Hassel began their first factory in Mathews's kitchen. Their kitchen table workshop became Zenith Radio Corporation, and later, Zenith Electronics Corporation. The Zenith Radio Corporation was formed in 1923 by former Naval Commander E.F. McDonald Jr., along with Mathews and Hassel, as the exclusive sales and marketing organization for radio equipment built by Chicago Radio Laboratory. When Zenith acquired the manufacturing facility in 1924, the company produced its first portable radio. The first home receivers operating on household AC current followed in 1926, and the first automatic pushbutton radio tuning was introduced in 1927. In that year, the slogan "The quality goes in before the name goes on" was first used.

On February 2, 1940, Zenith formed W9XEN (WEFM), the first FM radio station in the Midwest. Work at the station in the late 1940s lead to the first high-fidelity FM broadcasts in the early 1950s, followed by Zenith's invention of the stereo FM radio broadcast system in 1961. The last Zenith radio was sold in 1982, and the company name was changed to Zenith Electronics Corporation in 1984 to better reflect their specialization in television and other consumer electronics and network systems. Today these include interactive television, digital cable, video dial tone, networking, and HDTV systems.

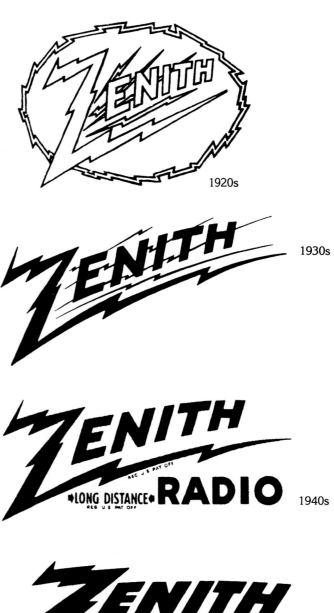

1920s

1930s

1940s

1950s

Selected Bibliography

Books

Breed, Robert F. *Collecting Transistor Novelty Radios*. Gas City, Indiana: L-W Books, 1990.

Bunis, Marty & Sue. *Collector's Guide to Antique Radios*. Paducah, Kentucky: Collector Books, 1991.

-------.*Collector's Guide to Transistor Radios*. Paducah, Kentucky: Collector Books, 1994.

-------. *Collector's Guide to Antique Radios*. fourth ed. Paducah, Kentucky: Collector Books, 1997.

Collins, Philip. *Radios: the Golden Age*. San Francisco: Chronicle, 1987.

-------. *Radios Redux: Listening in Style*. San Francisco: Chronicle, 1991.

Grinder, Robert E. *The Radio Collector's Directory and Price Guide 1921-1965*. second ed. Sonoran Publishing, LLC, 1995.

Hawes, Robert with Gad Sassower. *Bakelite Radios*. Edison, New Jersey: Chartwell, 1996.

Howard W. Sams & Co. *Radios of the Baby Boom Era 1946 to 1960*. volumes 1 through 6. Indianapolis: PROMPT Pub., 1991.'

Lane, David R. and Robert A. *Transistor Radios: A Collector''s Encyclopedia and Price Guide*. Radnor, Pennsylvania: Wallace-Homestead, 1994.

Sideli, John. *Classic Plastic Radios of the 1930s and 1940s*. New York: E.P. Dutton, 1990.

Stein, Mark V. *Machine Age to Jet Age: Radiomania's Guide to Tabletop Radios 1933-1959*. Baltimore: Radiomania, 1994.

Wood, Scott, ed. *Evolution of the Radio*. Gas City, Indiana: L-W Books, 1991.

-------. *Evolution of the Radio Volume 2*. Gas City, Indiana: L-W Books, 1993

Articles and other sources:

Motorola, Inc. "A Timeline of Motorola History." Schaumburg, Illinois: Motorola Museum of Electronics, 1995.

Zenith Electronics Corporation. *The Zenith Story: A History from 1918 to 1954*. pamphlet. Glenview, Illinois: Zenith, 1955.

-------. Zenith Electronics Corporation. *The Zenith Log*. Glenview, Illinois: Zenith, 1988.

-------. "Zenith Electronics Corporation." typescript. Summer 1995.

Index

Admiral,
*17, 35, 94,
98, 103, 122,
132, 190*

Alco,
113, 191

Arvin,
*31, 41, 60-
61, 69, 74,
81, 97,
102, 178,
185, 191*

Aircastle,
16

Ambassador,
24, 190

Belmont,
23

Airline,
*13, 81, 95,
191*

Arkay,
31

Bendix,
17, 28, 192

Bulova,
*90-91, 144,
151, 192*

Clarion,
30, 60

Delco,
15, 17

Capehart,
79, 145, 192

Classic,
114

Detrola,
10, 21, 195

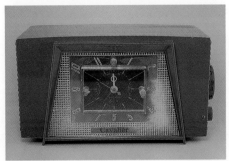

Cavalier,
132

Coronado,
69

Dewald,
32

Channel
Master,
117-118, 192

Crosley,
*27, 30, 48-57,
70-71, 76-77,
81, 134-135,
144, 180-181,
187, 193-194*

Dumont,
89, 195

Emerson,
*18-19, 36, 40,
42, 60, 82-85,
94, 99-100,
109, 111,
141, 147,
155, 178-179,
195-196*

Galaxie,
114, 197

Golden Bell,
114, 198

Fada,
*6, 13-14, 24,
196*

Garod,
31, 197

Hallicrafters,
185, 199

Farnsworth,
36

General
Electric,
*16, 40, 89,
107, 112,
124, 126-127,
129, 133,
136-137, 152
154, 156-159,
164-165, 183,
186, 197*

Jewel,
18, 124, 199

Firestone,
*21, 102, 147,
155, 197*

Gilfillan,
16

Majestic,
10, 12, 199

216

Major,
116

Mirror-tone,
26

Nipco,
189

Meck,
26

Monarch,
113, 200

Olympic,
31, 72-73

Merc-Radio,
11, 123, 199

Monitorradio,
110

Omscolite,
113, 201

Meteor,
97

Motorola,
*33, 75, 86-87,
92, 94, 98,
104-106, 109,
142, 147,
170-173, 189,
200*

Packard Bell,
40

Philco,
20-21, 30,
45-47, 68-69,
80, 92-93,
109, 182,
201

Sentinel,
24, 36, 81,
203

Sterling Delux,
68

Philharmonic,
124

Setchell-
Carlson,
24

Stewart,
89

Raleigh
Electronic
Corp.,
122

Silvertone,
17, 22, 37,
63, 101, 108,
119, 141,
160-163, 204

Stewart Warner,
16, 26, 132,
204

RCA,
9, 13, 41, 43,
69, 94, 120-
121, 174-
175, 179,
183, 186,
189, 202

Sparton,
130-131, 204

Stromberg-
Carlson,
13, 146

Sylvania,
*68, 97, 108,
115, 128,
155, 178 185,
205*

Truetone
(Western Auto),
*29, 36, 44,
108, 208*

Tele-tone,
*58-59, 63,
141, 206*

Westinghouse,
*30, 40, 84,
96, 109, 118,
124, 143, 208*

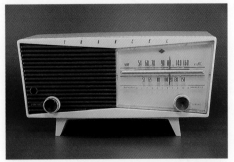

Trancel,
108

Zenith,
*8, 10-12, 20,
34, 38-39, 66-
67, 78, 88-89,
109, 125, 138-
140, 148-150,
162, 166-169,
181, 184, 186,
188, 209-210*

Trav-ler,
*26, 64-65,
78, 144,
176-177,
181, 206-
207*

Zephyr,
114, 210